The Cow Book

Marc Gallant

Alfred A. Knopf, New York, 1983

THIS IS A BORZOI BOOK

PUBLISHED BY ALFRED A. KNOPF, INC.

Library of Congress Cataloging in Publication Data
Gallant, Marc.
The cow book.
Summary: Presents a cow's eye view of the history
of civilization. Illustrated with paintings by artists
from around the world.
1. Cows—Anecdotes, facetiae, satire, etc.
[1. Cows—Wit and humor] I. Title.
PN6231.C24G34   1983   82-47816
ISBN 0-394-52034-3

Manufactured in Italy
First American Edition

For Mom and Dad

CONTENTS

The Cow Book

The First Four Million Years

DESPERATELY slow. Cows consider this period the least interesting in bovine history. The mighty aurochs (species *primigenius,* genus *Bos,* family Bovidae, order Artiodactyla), hollow-horned ancestor to the common cow, spends all of this period roaming about the steppes of Asia. Unlike today's cow, the aurochs is formidable; when she charges, dust clouds billow and lesser beasts move aside. The aurochs or wild cow is happiest, however, living in the background—she prefers to amble among leafy greens and keep an eye on things.

Precious little to watch until the great ice comes. Then a variety of two-legged creatures, all with thick skulls and heavy brows, make their way to the northern grasslands. Initially these hunters spend most of their time chasing wild animals and hurling stones into the air; they also spend a lot of time trying to keep their fires lit.

Things pick up considerably towards the end of this period. Weather worsens.

35000 B.C.–8000 B.C.

The last phase of the great Ice Age is now at its height. Extreme ice conditions are beginning to work on everyone's nerves.

Cro-Magnon man replaces the heavy-browed Neanderthalers. Wild cows are forced to reconsider these new advanced hunters; they have smoother foreheads and seem to be getting taller.

The "bull-roarer" is invented. By rapidly twirling a simple device over their heads Cro-Magnon hunters are able to produce a sound remarkably similar to the bellowing of a wild bull. Cows could easily do without this—bulls make more than enough noise as it is.

Ice conditions become almost unbearable. Cro-Magnon hunters head west. A subgroup, the Magdalenians, are inspired by the inclement weather to stay inside more and take up painting. The Magdalenians' first pictures are of wild cows, painted on the walls of caves. Cows are jubilant but not quite sure what it all means.

Hunters improve their spear throwing and invent the bow and arrow. In fact, cows notice that almost everywhere crude bits of flint are taking on new shapes.

The stockier bison (*Bos bonasus*), a cousin to the cow, makes its way across a great land bridge to a new world in the east.

About two thousand years before this period ends, the great ice sheets start to melt; there's a strong feeling of spring on the planet earth. Early man becomes less dependent on big-game hunting and turns to smaller animals; he also starts eating more wild vegetables. Cows think this is an excellent idea.

With the new spring, people are coming out of their caves. Simple tents of skins and bones are being built. Nothing exciting, but it's a start. The sweet glorious sound of a simple flute is heard and there's dancing around the fire. (To cows this is almost as exciting as having their pictures painted.) There is a clustering of tents and a sense of family.

Soon dogs, sheep, and then goats are seen hanging around the camps of early man.

8000 B.C.–5000 B.C.

If cows ever need to be reminded how important they really are, this is the period they think about—or at least this is *one* of them. In these next three thousand years man makes the big move—he sets the foundation for a totally new kind of life, and cows are part of it. The first real farms appear in uplands near the eastern end of the Mediterranean Sea.

In no time cows are sharing barnyards with two distant cousins. They have never felt particularly close to either sheep or goats, but all three belong to the same family. They're all Bovidaes.

Aside from putting cows in a better position to view things up close, domestication has some wonderful side effects. Being on the inside, under man's protection, they can finally shed their heavy horns; and while the process can be awfully slow, some are even getting new coats. In the wilds, it was always necessary to be more conservative— duller earth tones for better blending with the landscape. Now it's all glorious spots!

The presence of cows in the first barnyards has a positive effect on early farmers; there is another great rush of creative energy. Simple tents are replaced with new mud brick homes. Not everyone gets one right away, of course. There is a vast improvement in all kinds of domestic utensils and in crafts. Becoming domesticated and getting new coats are certainly wonderful things, but the idea of living in a town brings even more excitement. One of the first towns ever built, Çatal Hüyük on the vast Konya plain of Anatolia (present-day Turkey), has practically six thousand people, many of them with their own cows. Being in a place with so many things happening at once is confusing at first, but cows grow to love it. This is their first urban experience.

Cows in Çatal Hüyük have no illusions about who comes first. It is duly noted that bulls are making a much stronger impression on local farmers. Bulls' heads, both real and carved in stone, are being placed in shrines, and as if that isn't enough, murals of them appear as well. There seems to be some kind of cult thing happening.

When the population increases at Çatal Hüyük people start to move away. The great thing is that cows are going with them. First Greece, and then Crete, and then finally too many places to mention. This is better than anything cows have ever imagined. Cows take their first trip in a boat.

At the end of this period, cows find themselves near the graceful Nile. Then even more travel—all along the coast of North Africa and down the horn as far as Kenya; over to the plains of Mesopotamia, to Transcaucasia, Turkmenia, and Afghanistan. The names alone are almost too much.

5000 B.C.–3000 B.C.

Cows by this point realize that farming is not just a fad. Hunting as a way of life is disappearing; the more settled life of agriculture and trading is spreading to the valleys of the Nile and the Tigris-Euphrates. There is a general feeling in the cow world that something major is about to happen.

Men begin to cooperate more, dig more ditches and irrigation canals, keep cows, and do more trading. The plow is developed, followed by the wheel. Cows respond to these inventions with only mild enthusiasm; soon they are pulling heavy sledges, carts, and plows over extremely rough ground. They accept this new role without complaint but do not enjoy being beasts of burden.

A new type of cow is bred in Mesopotamia; by selecting only certain animals from their longhorn breeding stock, farmers in Elam are able

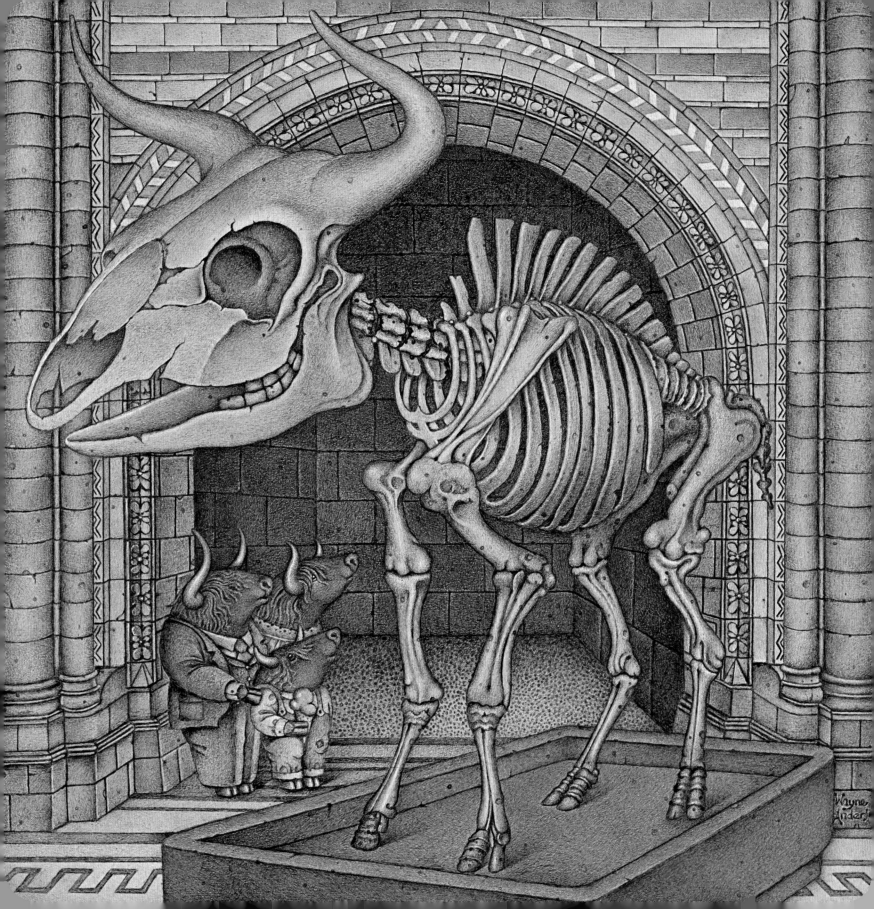

to produce cows with shorter horns. The new shorthorns (*Bos taurus brachyceros*) are equally admired for their narrow foreheads and long slender legs.

Men are now using copper for things other than trinkets and beads; they are learning to smelt the ore and cast it in open molds. Egyptians build real seagoing ships and invent the square sail.

For the first time cows notice small groups of men and sometimes even one man telling masses of people what to do. These men are called leaders; they seem to have convinced the masses that they speak for the gods or, in some cases, are gods themselves and must be obeyed. The leaders are served by a group of men called priests who spend most of their time around temples trying to keep the books straight. How much grain in the bin? How many cows?

As cities grow and life becomes more complicated there is a greater need to keep records. Priests invent a system of writing but keep the art pretty much to themselves.

A new musical instrument appears in the land of Sumer; as far as cows are concerned, the lyre, particularly when seen without strings, bears an uncanny resemblance to cattle horns. It's highly doubtful that anyone will notice, but cows are pleased just the same. A royal lyre is commissioned in the city of Ur; it is decorated not only with lapis lazuli but with a golden bull's head.

Farming spreads to India, Baluchistan, Sind. Cows are used as articles of trade between the new civilizations and the farmers in the east. The humped zebu (*Bos indicus*) is now seen frequently in the streets of Ur. *Bos primigenius* find the humps a bit off-putting at first, but admit that zebus have a certain elegance. The fact that they grunt instead of bellow doesn't help. Farming is now established in China. Cows find this part of the world interesting but would prefer fewer dogs.

More paintings of cows appear, this time on large rocks in the Sahara. Wild cows in Africa find this reassuring.

3000 B.C.–2000 B.C.

There's lots happening in this period, but cows spend most of their time thinking about Hathor, an elegant, horned cow goddess in the Nile Valley. They swoon at the mention of her name.

Hathor started out as a local goddess but now occupies a place of prominence in the Egyptian pantheon. She's a mother goddess, a goddess of love and joy, and a goddess of two things very special to cows—music and dance.

Hathor is a protectress of infants. Egyptians believe that she and her assistants are present at the births of pharaohs. Cows are uncertain about this but think it's a nice idea.

Life in the river valleys is reaching new heights, what people refer to as high civilization. There's a lot more hammering going on—construction everywhere. Zoser, second pharaoh of Dynasty III, asks his architect Imhotep to build him a step pyramid at Saqqara; before long there are pyramids at Giza. Egyptian royalty apparently believe they're coming back, and as they don't want to start all over again, they bury their favorite things with them under these huge mounds of stone.

Far away on the flood plains of the Indus Valley, another civilization is reaching its peak. Zebu cows are kept busy pulling cartloads of grain to the cities of Mohenjo-Daro and Harappa. The streets here are well laid out and there's even underground plumbing, but the civilization doesn't last.

The idea of making pottery on a wheel catches on in the Far East. Man discovers that by adding tin to copper you get bronze and much better weapons. There's a lot of talk in Mesopotamia

about someone named Gilgamesh. Towns are popping up in mainland Greece.

In Mesopotamia cows are still being milked from behind. People seem to think that if goats can be milked in this fashion, why not their larger cousins? To be truthful, cows would much prefer being milked from the side.

2000 B.C.–1000 B.C.

As this period opens, cows see the first signs of a complicated future. Life in the Near East is now fraught with turmoil; kings squabble over trade routes, armies plunder, and entire kingdoms come and go before anyone can have a good look.

Cows witness the creation of high civilization in Europe. On the island of Crete a group of people called Minoans build beautiful palaces and quickly establish a preference for doing things differently. At the palaces in Knossos, Minoans even place their pillars upside down with the wide part at the top. Cows are particularly fond of this culture and are wildly enthusiastic about their art; it's not nearly so pompous or monumental as art in the river valleys and it's much easier to look at.

Man's preoccupation with the bull continues. There's a strange Minoan rumor about a beastly man-bull named Minotaur who lives in a Cretan maze and has an appetite for human flesh. The Minoans also foolishly take part in a sport called bull leaping. Cows think this is highly dangerous. This is not a new sport, of course—the same sort of thing was going on in Çatal Hüyük four thousand years ago. There it was common practice to have one person hold the bull by his tongue while another jumped over his back.

Just when things are going really well on Crete, an earthquake destroys the palaces at Knossos; a few centuries later a volcano destroys the entire Minoan fleet. Europe's first civilization comes to an end.

At about the time that iron is replacing bronze, cows witness the emergence of a new people called Hebrews. They're following a man named Abraham who apparently has been promised a land in the north by the tribal god Yahweh. Cows notice that Abraham's people like to wander; they also keep many cows and quarrel with their neighbors over grazing rights.

Small numbers of shorthorn cattle are introduced into Egypt from the ancient land of Elam; at first the shorthorns are considerably outnumbered by their longhorn sisters who arrived here twenty-five centuries earlier, but local farmers are so impressed with the shorter horns that they begin almost immediately to produce a new variety of cattle, with no horns at all!

Semitic peoples migrating from Asia to the Horn of Africa take along large numbers of *Bos indicus* cattle. The humped Asiatic zebus wish they could have come to this part of the world sooner. Cows that have remained in Egypt watch quietly as the Pharaonic age ends. They are convinced that early Egyptians have spent entirely too much time thinking about life after death.

In the Far East an Oriental tribe called the Shang are busy creating China's first civilization. The Shang are masters in the art of making bronze and have the nasty habit of sacrificing humans and animals when they entomb their kings.

In India the Hindu god Indra makes a major proclamation: "Never forget the debt you owe to cattle. To you they are sacred." Indian cows are unsure about being sacred, but it's nice to be appreciated.

In northern India cows are impressed with a story about a sacred bull called Nandi. When a wealthy man dies, his relatives select the best bull they can find and dedicate it to Nandi. A

Brahman priest then brands the bull and sets it free to wander anywhere it wants; it is even permitted to eat cultivated crops without fear of punishment or banishment.

1000 B.C.–500 B.C.

The first Hebrew kings appear: Saul, David, and Solomon. Cows witness great technical advances and watch the building of a new temple in Jerusalem. Hebrew prophets introduce the notion of one universal God. This is not too popular at first. The prophets also question the whole business of special privileges in the priestly caste. Cows are pleased when this happens.

In the Far East the Shang have been replaced by the Chou, who are even less interested in the outside world than their predecessors. The Chou are convinced they're surrounded by barbarians. Cows think they might be right.

At the beginning of the Chou dynasty, major decisions are made by consulting oracle bones. Cows are usually not interested in this kind of thing but are tickled to learn that oracle bones often come from the bovine family.

Back in the river valleys there's more fighting than ever, especially among the Assyrians in northern Mesopotamia. Ashurnasirpal, an early king, builds a palace at Calah and to celebrate its opening gives a ten-day party for sixty-nine thousand five hundred and seventy-four people. To feed his guests, eighteen thousand bovines are slaughtered, cows as well as bulls, along with thirty-five thousand birds. Cows consider the party to be extravagant and in very poor taste.

In the west, Greeks and Phoenicians are busy creating new colonies. Greek colonizers introduce wine drinking to the barbarous Celts. Homer, a bearded Greek poet, writes a number of stories which catch on immediately. They're called epics.

Coins are invented by the Lydians on the east coast of the Mediterranean. Cows have rarely seen an idea spread so rapidly. In a very short time these new pieces of metal are in use throughout the civilized world. Money is the rage.

The Greeks come up with three new columns —the Doric, the Ionic, and the Corinthian. From the beginning cows find it difficult to remember which is which.

This period produces great thinkers all the way from China to the Italian peninsula. Confucius, Lao-tsu, Gautama Buddha, Zoroaster, Thales, Anaximander, and Pythagoras.

Zoroaster, the religious teacher from Persia, is of particular interest to cows; he teaches that bovines were the first animals created by the god Ahura Mazda, that life carried out in a certain way will be rewarded by an eternal afterlife. This will happen on a day of judgment. Cows hear the first mention of angels and hellfire.

There is talk about a new king living at the edge of the Tigris Valley. Cyrus the Great creates the biggest empire anyone has ever seen. He is a man to be admired. He frees thousands of captives from the city of Babylon and rebuilds the temple in Jerusalem; most important, under his guidance the Persian empire does a reasonable job of maintaining peace. This is definitely a first.

500 B.C.–A.D. 1

Cows take a look at the barbarous Celts, a group of tall, fair-skinned warriors who are forever attacking their neighbors or fighting among themselves. The Celts are partial to wine drinking and wearing jewelry, and they make a lot of fuss over horses. Celtic art is influenced by Scythian animal designs and by decorative art

on Greek and Roman wine vessels. Many of the Celtic tribes are known to keep herds of cattle; being a Celtic cow is not always fun, however— along with human beings, cows are sometimes part of a savage sacrifice made to the gods.

Life continues to flourish around the Mediterranean, particularly among the Greeks. Pericles, an Athenian statesman, builds the Parthenon and introduces a new form of government which allows people to choose their own leaders. Cows see democracy as a grand concept but refuse to get too excited; in practice these larger ideas don't always work out.

The Greeks produce some very creative thinkers, including Plato, Aristotle, and Socrates. They also build a temple to Zeus, open a lovely new concert hall, and perfect the art of vase painting. In the capital city of Athens cows notice that sculpture is becoming much more natural-looking; statues of humans appear relaxed and are often depicted with glorious smiles.

In the second half of the fourth century cows witness the creation of yet another empire, this one even bigger than the last. Led by Philip II of Macedonia, a group of rough characters decide to take advantage of the Persians, who are beginning to weaken. When Philip is assassinated, his conquests are continued by his son Alexander.

Cows have never seen the likes of Alexander the Great. For one thing, he never stops; in just ten years he defeats the Persians, claims the throne, and spreads Hellenistic culture all the way to the Punjab. Following his premature death, the Macedonian Empire is divided up among five of Alexander's generals.

In the Far East, China is united under one emperor. A great wall is built, irrigation projects are carried out, and the population increases.

There's yet another war, this time between Romans and Carthaginians. Cows in the Alps witness one of the strangest spectacles they've ever seen when Hannibal, a Carthaginian general, marches through the mountains with a herd of elephants.

The name Julius Caesar is heard frequently in this period. Cows in Rome are adversely affected by one of Caesar's first acts. In an attempt to solve the city's traffic problems, the new emperor forbids the use of all wheeled vehicles in the streets of Rome, except at night. Now, with increased street noise keeping local people from their sleep, cows in the transport business often bear the brunt of some very strong language.

Cows notice at least one exception to Caesar's new traffic laws: wheeled vehicles employed in the building industry are still permitted to travel in daylight. With dishonest contractors and profiteering landlords busy erecting flimsy tenement buildings, cows suspect some type of special arrangement has been made with the emperor.

Caesar's military conquests attract a lot of attention, but there's even more interest in his trip to Egypt. He meets a seductive young woman named Cleopatra and leaves her pregnant with a son. The Egyptian queen is said to improve her complexion by taking milk baths.

When Caesar is murdered and replaced by a threesome, the name Cleopatra is heard once again. This time Mark Antony meets Cleopatra and leaves her with twins. Later, Antony stabs himself and dies in Cleopatra's arms, only to be followed by the queen herself, who dies of a snake bite.

Egypt becomes a province of Rome.

Just when the Roman Empire is reaching a summit of prosperity, cows witness a major event in the Near East. Jesus of Nazareth is born in a tiny manger. Although cows don't immediately recognize the significance of this birth, they're more than willing to make their manger available.

A.D. 1–500

Life is getting extremely busy. Cows decide it would be better for everyone if they kept a little distance from people. There is a new cow stance: they appear less frisky, with heads slightly lowered; their entire demeanor suggests quiet aloofness. People will undoubtedly misinterpret this change in posture, but then again, if history holds true, they'll probably not even notice.

While cows are making these adjustments, the Chinese economy is faltering; the Romans are building a theater on recently annexed land in North Africa; and Jesus of Nazareth is nailed to a cross. It often seems that people with the most to offer are the first to go.

The Romans invade Britain and begin building a vast empire. A man named Saul does a lot of traveling and preaching among the poor. Cows notice that Jesus' teachings are especially popular with the poor; the number of converts seems to increase when the economy worsens.

There is a great deal of writing in this period. Ovid, a Roman poet, writes an impressive amount of poetry before being sent into exile for allegedly fooling around with the emperor's daughter. Tacitus, Juvenal, and Martial, all Romans, have a lot to write about in this period, most of it critical of the contemporary scene.

As the second century opens, the Roman Empire is at its height. Jews from Judea are forced to disperse when the Romans take over their capital. A Latin version of the Bible appears in Rome, and cows witness a revival in Stoicism. Cows find it difficult to get a clear picture of what Stoicism is about; one of its beliefs seems to be that man should live in harmony with nature. Cows could not agree more.

Cows realize that anyone arriving in the city of Rome for the first time could easily be fooled. On the surface the Roman capital projects a formal dignity. It is only by looking much more closely that one can see what this city is truly like.

An immense amphitheater is built in the city of Rome. To celebrate, large numbers of human beings and countless numbers of wild animals are slain; in one day alone, five thousand animals are killed. Tigers, wolves, deer, antelope, wild horses, leopards—creatures of all descriptions are set against each other in battle. Entire herds of domestic cattle are thrown into the arena and slaughtered just to keep things going—this is called padding. Cattle are also required to haul cartloads of bodies to the Roman carnarium, an assemblage of open pits which line the city wall.

Nero becomes Roman emperor. Although his reign starts out with good intentions, cows come to view this man as a symbol of Roman madness and depravity. After murdering his own mother and setting fire to Rome, Nero begins a systematic slaughter of Christians. Next, he executes many of his opponents after discovering their plot to get rid of him. When revolts break out in the provinces and Nero's own guards turn against him, he is forced to flee the capital. Cows hear shortly after that the emperor has taken his own life.

In Ireland cows are indirectly involved in the selection of a new king; this happens when the Celtic clans assemble on the hills of Tara for a ceremony called *tardfies*; this means "bull dream." The ceremony begins with the ritual slaying of a white bull; the bull is then consumed by the chief priest or druid, causing him to enter a powerful trance. It is during this trance that the new king's name is revealed.

This reminds cows that as far back as the early Chinese, people depended on bovines for direction. Remember the Chou use of bovine oracle bones?

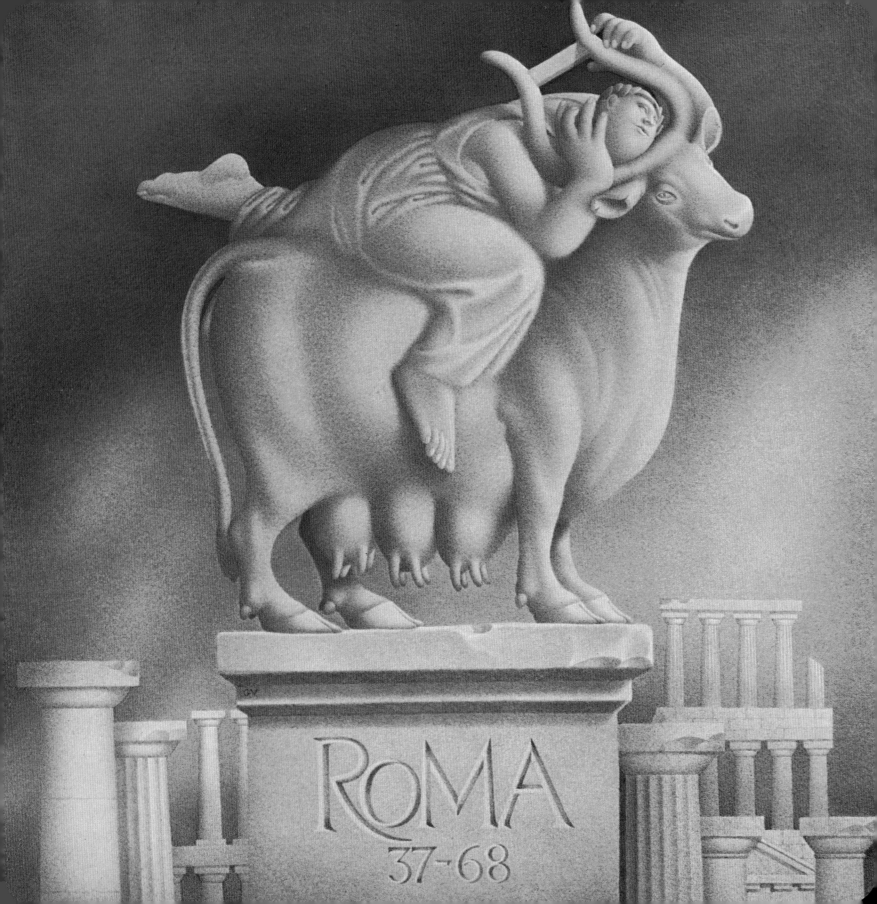

Cows in Scotland witness another example of dependence on bovines; the Scottish Celts take the hide from a freshly slain bull and place it, fat side up, on round wattles made from the quicken tree. After a night of sleeping on this hide, the seer makes a prophesy according to his dreams.

There is great excitement in India when Brahman priests instruct the Hindu population to venerate the cow. Henceforth people are expected to treat cows with respect and are forbidden to feed on them. Cows are somewhat confused by this sudden adulation, but in time come to understand that cow worship is directly related to Ahimsa, the Hindu belief in the unity of all life.

Devotion to the mother cow takes on numerous forms, including the practice of decorating the sacred animals with golden handprints. Cows welcome this kind of attention even if it only happens on special occasions.

At the beginning of the third century there is talk about a code of Jewish laws called the Mishnah. As cows see it, even if Jews can no longer live together in one place, they're expected to live according to these laws; the Mishnah will help make them feel they still belong to the same family.

In the Roman Empire, Emperor Caracalla is building an immense public bath. For some time now Romans have had a thing about bathing; they seem terribly interested in their own bodies. Bovines of course are expected to haul cartloads of wood from the country to keep the bath water hot. The emperor also gives citizenship to most free inhabitants of the empire—a dubious honor, as far as cows are concerned.

There is a lot of confusion among the military. Of the twenty-six emperors in this century, eighteen are brutally murdered by their own troops or assassinated by the Praetorian Guard; two are killed in battle; one dies in captivity, another from the plague. Two are smart enough to abdicate, and two die naturally. Little wonder cows don't feel comfortable among the Romans.

By mid-century, German tribes have begun moving into the empire, and a Greek living in Alexandria invents algebra—very few people are excited about algebra at first. Towards the end of the century Emperor Diocletian decides to divide the empire into two parts, east and west, with an emperor and assistant ruling in each.

The Asian foot stirrup is developed at the beginning of the fourth century. Soon nomads mounted on horseback are charging about in every direction. Huns pour into northern China, causing vast numbers of people to move south; they establish countless dynasties but are so inept at running a country that the Chinese are required to help. The Huns become so Chinese-looking that cows can hardly tell them apart.

In the Western world Christians continue to be persecuted and more baths are being built. An Egyptian hermit, Anthony, has just come back from the desert with the idea of living in a monastery. Anthony is interested in self-denial—cows think a bit of self-denial would be good for everyone. A new emperor, Constantine, becomes a Christian and builds a city in his own name.

While the new Roman capital at Constantinople is taking shape, an Indian leader is creating the Gupta Empire not far from the ancient city of Mohenjo-Daro.

During the fourth century a bishop from Alexandria introduces Anthony's monastery idea into western Europe; before long a German tribe, the Goths, convert to Arian Christianity.

In the Far East the Chinese are drinking tea; the Japanese are taking an interest in Korea, learning to write in Chinese, and exploring the concepts of Confucianism. Towards central Asia, the Persians are flourishing under Shapur II.

B. G. Sharma
UDAIPUR (INDIA)

Cows think of the fifth century as a very trying time. People in the western half of the Roman Empire appear to be in a slump; no one has any gumption. Occasionally a crumbling wall may get repaired, but more often than not it will be left. When things do get repaired it's not unusual to see people using bricks and stones from collapsing public monuments.

Farmers and tradesmen, along with people who can't do much of anything, continue to settle inside Roman camps. Cows notice that as soon as these camps reach a particular size and become fortified, the bishops move in; this way, when the empire begins to falter and local officials lose their power, the bishops are there to take over. This is not a particularly good time to inherit a town however. The last Roman emperor in the west is finally deposed. Towns and cities everywhere are in a terrible state.

Aside from dealing with crumbling society, people in the western empire are forced to cope with poor crops and an endless flow of raiders pouring in from both the north and the south. Vandals, Alani, and Suevi ravage Gaul and Spain; Visigoths invade Italy and sack Rome; tribes living in the Low Countries are forced to move west; suddenly cows who have been living quietly among the Frisians are squeezed into narrow boats and taken off to an island province—Britain. Life here is not dramatically different—the British are fighting people from the north called Picts.

Amid all the confusion, a new kingdom is established in Europe. A Germanic people led by Clovis invade Roman Gaul. The Franks are more willing than most tribes to settle down but they're definitely not impressed with Roman-style government. Towns decay even further.

The Ostrogoths take power in Italy. In Britain the local population manages to enjoy some peace and quiet after beating back a band of Saxons.

Sixth Century

Monasteries are popping up all over Ireland—they're even placed on tiny obscure islands. Cows in Ireland become involved in a legal decision over who has the rights to a particular book. Columban, an Irish monk, loans a manuscript to a fellow churchman and then discovers a copy has been made. When Columban insists that both the manuscript and the copy are rightfully his and must be returned, a terrible battle takes place. The Battle of the Book is finally settled when a decision is reached by the high king himself. He rules as follows: "To every cow its calf, and to every book its copy." Cows view this ruling as fair and proper.

In Italy, Benedict and a few monks establish a monastery at Monte Cassino. Justinian becomes emperor in the east and sets about trying to recover the western provinces. With the provinces in such awful shape, cows wonder if it's worth all the trouble.

By mid-century a third of the west has been reconquered and Justinian has completed a magnificent domed cathedral in Constantinople. Cows have rarely seen anything more dazzling than the cathedral of St. Sophia; they are shocked and disappointed when after only twenty years, the wonderful dome collapses. The emperor of course has it repaired; only this time he makes sure that buttresses are used.

The nomadic Turks are starting to move about. This forces the White Huns into Europe. En route the Huns join up with other tribes and call themselves Avars. This new and larger group push the German Lombards into Italy. Before long, Lombards have taken over much of Justinian's reclaimed territory.

Aside from the plague, not much more happens in Europe.

Seventh Century

In the desert land of Arabia there's great excitement when an elephant is brought over from Africa to assist in an attack on the Kabaa temple. In the same year, Mohammed is born. The prophet Mohammed preaches submission to the will of God. At first Mohammed's preaching does little more than upset his family, but with help from Bedouin herdsmen the prophet enters the city of Mecca, destroys the pagan idols, and brings Islam to the world. Convinced that death on the battlefield leads to paradise, Arab armies go on the rampage. Syria, Palestine, Mesopotamia, Egypt, and all of west Persia are captured in Allah's name. Arabs fight with Berbers in the north of Africa and ride into Spain.

In China the Sui reunite the country and begin the biggest project cows have seen in years. A canal connecting the Yellow River to the Yangtze requires five and a half million forced laborers. Cattle, of course, as beasts of burden are not far away. The Sui are replaced by the T'ang; friendly relationships are established with neighbors in the west. Chinese philosopher Hsüan-tsang travels to India; cows are pleased to see the Chinese getting out more.

In Japan, the ruler Shōtoku is calling for centralized government. The Japanese are so impressed with Chinese culture they spend much of this century trying to emulate it. Cows notice monasteries are almost as popular here as they are in Europe, but the monks are Buddhists.

Eighth and Ninth Centuries

As the eighth century opens, Chinese poets Li Po and Tu Fu are basking in popularity. Within thirty years the Chinese are printing on paper.

There is a second wave of Arab expansion, this time to the east. When people in this area turn to the Chinese for help, the Arabs fight them as well. In mid-century abu-al-Abbas is proclaimed caliph. His Abbassid empire is smaller than the previous one, but its court is luxurious and the Abbassids are known for their interest in learning: at one point they come up with something called trigonometry. Cows feel that Abbassids spend far too much time building elaborate hunting lodges and chasing hawks in the desert.

As a direct result of crossbreeding, African herdsmen have developed a bewildering assortment of exotic cows. The cattle herders are fond of big or unusual horns and pay considerable attention to coat color and markings. Cows, for their part, are intrigued by the Africans' use of decoration both on themselves and on their livestock. Some of the tribes cut narrow strips of hide from the cow's neck to form a fringe—this is meant to identify animals with their owners. Cows would prefer a more colorful approach; maybe someday they'll even get to wear tribal jewelry themselves.

Back in Britain, Offa, king of Mercia, is building a huge dyke along the entire Welsh frontier; members of the bovine family do a lot of the heavy work. On the continent, among the Franks a young man named Charles succeeds Pepin the Short as king. Charles chases the Lombards, conquers the Saxons, and destroys the Avar kingdom. Cows are not at all surprised when Charles is proclaimed Emperor of the West.

Rugged characters in the far north learn to make a really good sailing ship; in the last decade of this century the Norse of Hordaland, sailing their new ships, discover the Shetland and Orkney islands; they also explore the coast of Britain and plunder the monasteries of Lindisfarne and Jarrow.

Harun al-Raschid, fifth Abbassid ruler, gives

away an African emirate and starts a craze. Eventually he's left with only Iraq and west Persia. Cows think Harun's been awfully silly but realize it doesn't matter—with all the hawk hunting and good times, the Abbassids are not destined for survival anyway.

In the north, Swedes explore the rivers of Russia; they establish trade links between northern Europe and the Near East and create the first Russian state. The Danes join the Norse and end up with a good part of England. Norse explorers, after pestering the Irish, discover an island to the west. On finding the weather in Iceland no worse than back home, they establish a base there.

The ninth century marks a low point in the life of European cities. Besides the Vikings, the Moslems are still acting up. Sacking and pillaging of cities is a popular pastime; even the bishops do their share.

Tenth Century

By the tenth century Europeans are getting smart: in response to all the looting and raiding, walls are going up around towns, castles, and monasteries. Many of the old Roman towns had foolishly allowed their walls to crumble, a classical example of people expecting someone else to take care of important matters.

A French king gives Rollo the Viking land in the north of France on condition that he stop attacking the kingdom. Cows realize by now that however rough these big Vikings are, they're really just looking for a place to settle down.

In the Near East, hunting days are over for the Abbassid. People claiming descent from Mohammed's daughter create a new dynasty and divine revelations as given to Mohammed are written down in the Koran.

At the very end of the century a Norwegian explorer is banished from Iceland for committing murder. Travelling west, Eric the Red discovers a large ice-covered island which he names Greenland. Cows are confused by the name but can hardly contain themselves when they realize there could be a whole new world out there.

Eleventh, Twelfth, and Thirteenth Centuries

The eleventh and twelfth centuries immediately appear brighter in Europe. The pieces of a broken-down empire have been picked up. There is a sense that humans with all their capacity for destruction are in the end more interested in survival.

Much of the impetus for this rebirth comes directly from the Church—most visibly from monks, who do much of the work. The practical monks of Saint Benedict, the reform-minded Cluniacs, and the back-to-the-land Cistercians are all in a flurry. Their monasteries not only inspire settlement but act as centers for learning, places where classical works from the past can be written down. Cows are delighted to play at least a minor role in this crucial act of preserving knowledge: often the ink used to copy the great works is stored in hollow bovine horns.

Church zealousness increases. A new skyline is created throughout the old empire. Christian architecture is everywhere. When the cathedral at Canterbury burns, rebuilding begins almost immediately.

In China the Sung dynasty is throttled by a tribe from the north. Cows watch the building of a huge funerary temple at Angkor Wat and hear the first big explosion ever near Nanking.

The thirteenth century gets off to a bad start when King John of England marries Isabella, a

French girl who is promised to someone else; this provokes an Anglo-French war. In the meantime Christian crusaders are marching to Palestine, although the Fourth Crusade seems to lose its focus. Instead of going straight to the Holy Land, the crusaders storm and sack Constantinople, all in the name of Jesus Christ.

Thousands of children form a Children's Crusade, heading out on foot to capture Jerusalem. Most of them perish along the way or get snapped up by the slave trade; none of them reaches the sacred city. This kind of thing can happen when people with lofty ideals become too zealous.

Pope Innocent III gives Francis of Assisi and eleven of his friends permission to roam about as preachers. The Franciscan order soon follows; cows can hardly keep up with these friars and monks. Of all of them, Francis of Assisi is the one cows would like to know best.

On the Siberian steppes a man named Temujin is proclaimed Genghis Khan, Very Mighty King. Cows pay as little attention as possible to this man, and when he manages to crush even the horrible Huns, they know for sure to stay away from him. When Genghis Khan dies he leaves his empire to a bad bunch. His sons Ogadai, Jagatai, and Tului and his grandson Batu continue to raid and pillage. Batu turns to Russia and manages to seize Moscow. Russian cows think about the name Moscow and decide that any resemblance to their own name is purely coincidental.

Finally, another grandson of Genghis becomes chief of the Mongols and ruler of a vast empire, including all of China. Fortunately, the Kublai Khan is a big improvement on his grandfather; he's open to other cultures and does a great deal to establish trade and communications with Europe. Much of this he does through a young Italian adventurer, Marco Polo. Cows cannot think about young Polo without wanting to go on about him; in short they think he's wonderful.

In the gentler arts, Islamic architecture spreads to India; universities appear at Cambridge, Padua, Toulouse, Lisbon, and Salamanca; a French poet writes a love poem called *Roman de la Rose;* and the Italian Dante writes a poem called *The Divine Comedy*—it's not very funny.

In the Nilgiri Hills of southern India cows can hardly believe the respect shown to the local dairyman. The dairy he works in is considered holy; the dairyman himself is received as a god. Everyone must bow before him, and no human except another dairyman may touch him. Cows think this might be taking matters a bit too far.

In the north of England wild white cattle are now expected to live within an enclosed park of approximately three hundred acres. There will be no more free range and no further contact with other breeds. This is not something Chillingham cattle are pleased about, but they are resolved to make the best of it.

The conflict between Church and State rages on, regardless of country. There's a lot of fighting over how much money should flow back to Rome and whether or not churches should be taxed locally. At the end of the century King Edward I of England outlaws the English clergy for refusing to pay their taxes. The practice of throwing important offenders out of the Church continues.

Fourteenth Century

Italian artist Giotto di Bondone paints frescoes in both Padua and Florence. His work is so appreciated among the Florentines, he's given a job as architect to the city. Also in Florence, a Greek scholar becomes the first man to lecture on classical culture in western Europe.

Another Italian living in the hills of Tuscany invents an exciting new dessert made from frozen milk. Cows are fairly certain this idea came to Bernardo Buontalenti by way of Marco Polo, who borrowed it from the Chinese.

In England, Geoffrey Chaucer is writing a book of tales, and the Gothic style of architecture is being tried out in a new choir building.

Another French Isabella is creating trouble for England. Married to Edward II, she leaves for her native France and returns to England with a lover who throws Edward into prison, where he dies. Edward is succeeded by his son Edward III, who reigns with Isabella's lover as regent. Three years later young Edward sends his mother to a nunnery and executes the lover. This sort of thing is happening all too often.

Cowbells are now in vogue throughout the Alpine region of central Europe. Metal cowbells not only help a farmer find his herd, they are also used to establish rank—the bigger the bell, the bossier the cow.

In the church, Clement V moves his headquarters from Rome to Avignon. When Urban VI moves the papal headquarters back to Rome, a rival pope is elected to stay behind at Avignon, starting a Great Schism in the western Church.

Jews are expelled from France and persecuted in Germany. The Statute of Kilkenny forbids English colonists to marry Irish citizens—this is expected to keep them loyal to the Crown. Cows expect it to create some real trouble.

Towns in northern Germany form a trading federation. Flemish weavers make a trade treaty with England. Around the same time, England and France start the Hundred Years' War—cows grow weary just thinking about it.

A decade after the war starts, bubonic plague sweeps through Europe, this time causing more human misery than ever; one third of the population of Europe plus thousands of domestic animals die from the Black Death and war. England is forced to bring in wage and price controls. The Dutch develop canal locks.

In Persia the lyric poet Hafiz writes so many love poems that few people read them all. Further east in Japan, a cultural revival is going on.

The great Mongol dynasty ends in total anarchy and is replaced by the Ming. Another Mongol, Timur the Lame, thinks of himself as a restorer of the empire and sets out to prove it. Finally, the Mongol holdings get broken up into khanates.

As the Mongols shatter, the kingdoms of Norway, Sweden, and Denmark unite under Eric of Pomerania.

Fifteenth and Sixteenth Centuries

Exotic travel is what cow dreams are made of—with one country after another striking out in search of new lands, cows in this period are often in a state of bliss. But it is not until late in the fifteenth century that a Spanish-backed Italian navigator has the good sense to include some giant-horned cattle on his New World voyage. Cows are in ecstasy!

Columbus makes his brilliant decision to bring along bovines, and cows queue up for a chance to be included. When the lucky ones finally arrive in Hispaniola, they have grown so fond of their Italian captain most of them refer to him as simply Christopher.

As part of his preparation for a trip to Mexico, Spanish conquistador Hernando Cortes selects a number of long-horned cattle from European stock living in the West Indies. If these bovines had known in advance that Cortes planned to launch an attack on the Aztec empire they would not have been nearly so anxious to cooperate.

In Mexico, Francisco Vásquez de Coronado, selects five hundred long-horned cattle as part of his entourage when he sets out to discover the golden cities of Cíbola. Mexican cows don't really expect young Coronado to find any gold, but they are quite willing to take part in his adventure.

In the meantime, two Portuguese explorers en route to India sight zebu-type cattle living among the Hottentot tribe in southern Africa. The first explorer establishes appropriate place names: Cow River, Cows Cape, Cape of the Cattle Herders. Local cows are pleased. The second explorer is so impressed with Hottentot cattle, he describes them in his journal as very big, very fat, and very wild. Cows could question the accuracy of this description but find it rather flattering.

Back in Europe the Church is in turmoil: at one point three popes are claiming two thrones. Life outside the Church is hardly better. In England Henry V becomes the first English monarch to read and write easily; cows think it's about time. Thousands of farmers from Kent march on London in the hopes of lowering their taxes. In France a zealous young peasant girl makes a name for herself before getting burned at the stake. The big event of the sixteenth century occurs when Ottoman Turks crush the eastern Empire—cows expected this much sooner. At the end of the century, cows watch helplessly as Jews are expelled once again from Spain.

A German priest takes exception to the selling of indulgences and ends up with a church of his own. Cows hear it said that Protestants intend to eliminate ungodliness from their church; they decide not to stop chewing their cud while this is coming about.

Pope Pius V, within months of his coronation, strongly condemns the gruesome sport of bullfighting; it soon becomes apparent that the pontiff's concern lies with the bullfighter, not with the mistreatment of bulls or horses. Cows are disappointed but hardly surprised when the papal condemnation fails to win a favorable response.

Bovine interest in music and art reaches a new high. Access to music is rarely a problem. If Italians are not singing madrigals, cow drovers are playing their flutes. Access to great paintings is not so easy but cows console themselves with the latest architecture and with good examples of outdoor sculpture; occasionally they catch a glimpse of a painting being carried from an artist's studio. Such is the case with Pieter Brueghel's painting of a Babylonian tower; cows think this painting is quite lovely but would have handled it differently—for some reason Brueghel's work reminds them of cheese.

Seventeenth and Eighteenth Centuries

Even a quick glance at the seventeenth and eighteenth centuries is enough to induce fatigue. If cows were not incredibly stubborn they would retire at this point, but bovine will and curiosity cannot be easily suppressed.

Business is big, cows have never seen such buying and selling: banks, banknotes, and adding machines all make their appearance. In the colonies, Manhattan Island is purchased by the Dutch for only sixty guilders—to cows this is the steal of the century.

While fiery young preachers are spreading gospel in America, many Europeans seem to be rejecting the Church entirely. In retrospect, cows are certain that Pope Urban, in the time of Galileo, made a grave error in not recognizing which ways the planets move. This is not the kind of mistake that inspires confidence.

In Lincolnshire, England, a mathematician named Isaac Newton develops a theory about

gravitation when he sees an apple falling from a tree. Cows suspect they should be impressed by Mr. Newton's intellect, but when he suggests that the earth was created in about 3500 B.C. they realize that he's not as smart as people say he is.

Cows in Britain are crowded into ships with shady-looking characters and sent off to Australia; this brings a sudden end to relative peace among the Aborigines. In the American colonies, the Revolutionary War brings an end to British interference. In France, the French Revolution brings an end to the monarchy until Napoleon Bonaparte brings an end to the revolution. Cows are grateful when all this is over.

In the south of Africa the Dutch East India Company acquires large herds of cattle from the local Hottentot tribe to feed thousands of sailors on their return trips to Europe as well as Dutch sailors en route to India. When the Hottentots discover that Europeans are taking advantage of them, they refuse to do any further business and force the Company to get into cattle breeding themselves. Before long, Hottentot cows are mating with bulls imported from Holland.

Cows living among the Zulu and Swazi tribes of southern Africa find themselves involved in an ancient custom which dictates that men should pay for their wives with cattle. When the bride is not of royal descent, thirteen animals are exchanged: one is presented to the bride's mother "to wipe away her tears," two are slaughtered and served at the wedding feast, and the rest become property of the bride's family. Senior tribesmen often take more than one wife, so Zulu and Swazi cattle herds are sometimes quite big.

Zebu cows in India are called into battle against the British. A small herd of humped zebu known as the Amrit Mahal becomes part of a military operation in the southern state of Mysore. Amrit Mahal cattle have been bred for warfare because of their fiery temperament and their ability to endure hardships. Now they are required to pull heavy military equipment to a war they have no interest in. This is not the first time cattle have been used in warfare: during the Middle Ages they were ridden into battle when horses were in short supply.

Cows pause to consider the common cat. These small whiskered animals are allowed to hang around farmers' barns in the hope that even smaller whiskered animals will be kept under control. Cows think it only fair that on occasion cats are rewarded with a taste of warm milk.

In America, a judge publishes the first strong protest against slavery. England and Scotland unite and the name Great Britain is heard for the first time—English cows are not sure if this means that Britain is now great or that it is great to have Scotland join them.

However important other events might appear, cows consider them boiled chaff next to the real revolution—a revolution in agriculture. Cows think of this period as the *Bovine Age of Great Improvement.* Farmers in the English Midlands develop the principles of selective breeding in cattle. Suddenly cows are being scrutinized: farmers, while fairly sure of their bulls, are often at odds over what makes the best cow. Crumpled horns and hairy dewlaps no longer seem important; science is de rigueur. From now on, cows will no longer be regarded as triple-purpose: good milkers will not have to pull plows; good draft and beef animals will not be expected to provide milk. To cows, this is enlightenment at its best.

Cows in remote areas of Britain are still on occasion being burned alive as sacrifices in an attempt to ward off cattle disease; one would think that such archaic practices would no longer exist, but people in remote areas are not always willing to give up their old customs.

Meanwhile cows in New York hear the first mention of ice cream being sold commercially—a Mr. Hall on Chatham Street has the good sense to introduce his "frozen delicacy" on an extremely hot day in June. Cows expect him to do well.

It soon becomes obvious that the Great Bovine Age is directly related to a revolution of the human mind called the Age of Reason or Enlightenment. Cows think this is by far the best kind of revolution: it's relatively quiet, not at all costly in human lives, and generally quite positive. Scientific inquiry and reason are replacing ignorance and superstition; academies of science, dance, music, and art are established; education becomes compulsory, first in Prussia, then in France; plants are classified; and printed books are now readily available.

Swiss cows living near Lake Geneva receive word that the illustrious Voltaire, French philosopher and man of letters, plans to purchase fifty cows for his new estate at Ferney. Cows have not been so excited since Columbus announced his decision to transport cattle to the New World.

British physician Edward Jenner makes a brilliant contribution to medical science when he discovers that people who've had cowpox have built-in protection against smallpox. His acute observation leads to the development of cowpox vaccine, an effective preventive treatment administered by inoculation. Cows are grateful to Dr. Jenner for his coining of the word vaccine. It is derived from *vacca,* the Latin for cow. This kind of direct recognition is not always forthcoming.

Off the French coast, on the island of Jersey, a decree is passed prohibiting the importation of cattle. Jersey cows realize this is not meant to stop them from mingling with foreigners; it is meant to stop French cows from island-hopping their way to lucrative English markets.

Nineteenth Century

The next hundred years mark the most exceptional period cows have ever witnessed. It is a century of promise and hope. Europeans enjoy a total of sixty-two years of virtual peace—this alone is enough to leave cows giddy with excitement. In the West, basic living conditions improve dramatically. Thousands of years of human stumbling seem to be over at last.

With steamships, passenger trains, automobiles, and airships all being developed, it's little wonder cows are taken with the nineteenth century. People who are not able to travel in this period can always send messages. In Britain, the penny post is established; in America, people are talking to each other through telegraphic wires. It's some time before cows figure out just how dot-and-dash messages can be sent through a wire, but they're very impressed with the concept.

Better ways of dealing with sewage and advances in medicine lead to the control of many diseases. Human populations increase at an extraordinary rate. Machines are performing duties which only a few years before had to be done by hand. Industrialization in the Western world is bringing about a greater concentration of wealth and power than anything cows have ever seen.

People rarely stop to consider it, but cows and agriculture are often at the heart of great revolutions: when hunters became farmers and true human advancement first began, cows were a substantial part of their early food supply; when plows were invented, who provided the muscle power to pull them? Cows do not often think about this because it tends to go to their heads, but the simple truth is—cows were at the center of the agricultural economy on which Western civilizations are based.

Although the nineteenth century is undeni-

ably great, cows are not suggesting that human creativity is suddenly sweeping the entire world or that misery is all in the past. Life outside Europe is not nearly so inspiring.

In China, while rebellion and famine claim thirty-five million lives, foreign powers are busy taking advantage of the country's weaknesses. The British are particularly enamored of China, where they have grown fond of a substance called opium; interest in the opium trade is so strong the British government goes to war to insure a constant supply.

The Russians are busy acquiring even more land: during this century they take all of northern Asia. Cows cannot imagine what they want with all this territory.

In India British involvement is far more extensive than cows had first realized. There is endless discussion about the betterment of Indian life, but aside from the creation of a new elite, cows see little progress. The British make a grievous error when they decide to use the fat of sacred cows to grease their military cartridges; this kind of insensitivity is not unusual—in fact, it's rather typical of European attitudes during this period.

European involvement in Africa is not a story cows are fond of telling. After years of slave trading, the major European powers seize the continent and partition it among themselves.

In Latin America, people seem to enjoy fighting: if they're not involved in open warfare, they're embroiled in hot political debate. Governments change so frequently cows can never be sure who's in power.

Cows on the European continent are hearing more and more talk of changes in British agriculture, but are starting to wonder if anything is ever going to happen for them. Serfdom and the old feudal system have been totally abolished, but no one would ever know it. In Germany, cows are forever seeing new pieces of farm machinery— every time they turn around there's another traveling exhibition—but for all the shiny new equipment, peasant life is not improving all that much.

In the British Isles, the Age of Bovine Improvement continues. Official recognition is given to a new type of cow in the county of Ayr. Neighboring cows are pleased, for the Ayrshires have not had an easy time, having been raised in an area where winter rations were slim and where they had to put up with some terribly wet weather.

Trying to keep up with local standards is not easy if you are a Swiss cow. Every year, following a difficult climb, Brown Swiss cows get to spend the entire summer with their heads in the clouds. Alpine cows are also beginning to enjoy better food—turnips, beets, and high-quality hay are now common fodder. The reason for this sudden change in diet is soon discovered—with recent improvements in the cheese industry, Brown Swiss are expected to provide extra milk.

In Britain, Thomas Cook starts a travel agency. Cows wonder if this man is any relation to Captain James Cook from the last century; Captain Cook had been interested in travel all his life.

Cows in America can be sure of one thing— Americans like to do things in a big way. During the first few years of this century they purchase a vast section of land from the French, an area almost four times as big as France itself. Other huge areas are annexed—taken by force or purchased with a threat of force. All of this new territory is to be explored and exploited. In a massive westward migration, thousands of zealous immigrants settle one area after another; during the process, countless tribes of Indians are either forced to abandon their homes or are destroyed entirely. This is the Wild West; cows are not always pleased to be a part of it.

Indians are not the only victims of nineteenth-century American expansionism: when settlers move westward a close cousin to the common cow almost disappears. In less than seventy years millions upon millions of wild buffalo are systematically slaughtered for both hides and meat. Often they are killed for their tongues only or shot down just for the fun of it, from the safety of fast-moving trains.

If anyone had suggested back in the first century that the Alexandrian notion of steam power would end up as a major force in the nineteenth century, cows would not have believed it. Now puffs of steam are streaming from tall factory chimneys and billowing from train-stacks and oceangoing ships. Steam is everywhere! An invention cows could have done without is the steam hammer.

During the first four decades of this century, a musket with interchangeable parts is designed; American engineer Samuel Colt patents a new type of revolver which becomes unusually popular; and the harmless Fulton steamboat becomes a steam warship. Cows are not impressed. They are even less impressed by a contraption being placed on the front end of trains; this iron grill is meant to clear railway tracks of cows and horses. It's called a cowcatcher. American bovines find this kind of aggressiveness quite unnecessary.

An American sailing ship on its way home from France stops briefly at the island of Guernsey. The ship's captain, a Mr. Prince from Boston, is so impressed with Guernsey milk he decides to purchase two young cows and one bull. Local cows are soon in a flurry. Within months, Guernseys are established on a small island in the distant state of New Hampshire; the winds in this part of the world are much cooler than gulf-stream breezes in Guernsey but Channel Island cows are famous for their ability to adapt.

In the sixth decade someone named Amelia Bloomer popularizes a new kind of personal garment for women; given weather conditions in America, cows think bloomers are extremely sensible.

Fighting breaks out in America when a number of states decide to leave the Union. Cows know that maintaining a strong union is of great concern, but the freeing of slaves is also an issue. The age-old practice of forced human labor is drawing to a close. At war's end the Union government is strengthened, and four million black slaves are set free—at least by law.

German philosopher Karl Marx is beginning to make a name for himself in Europe by openly suggesting that the world should be restructured as a classless society in which individuals can develop fully in a spirit of sharing and cooperation. Cows would dearly love to see such a world but do not think it will come about soon.

Charles Darwin publishes a book about the origin of species. The fuss this creates is hard to believe. Who in their right minds can deny that things evolve? People should be grateful: if it wasn't for evolution they would still be walking about on their knuckles and living in drafty tents.

In northern France cows are intrigued by the experiments of a dedicated scientist named Louis Pasteur. His work leads to the discovery that abnormal fermentation in beers and wines can be prevented by simple heating; a sigh of relief is heard in the wine industry. Soon cows see the same method being applied to milk: heating it to a certain temperature for a prescribed time kills harmful bacteria without damaging the milk's protein. The process is called pasteurization. If cows were in a position to pass out awards, Louis Pasteur would be among the first to get one.

A Scottish speech teacher working in America patents the telephone. Cows rarely see this

11 Juin 1983. C Lapointe

invention up close but recognize it as a major contribution: they are solidly behind anything which helps people communicate. One curious aspect of these new telephones is that when people are on them they tend to speak much louder than normal; suddenly cows in pastures are hearing one side of many conversations they would never have heard before the telephone arrived. They enjoy trying to figure out what's being said on the other end.

The American Guernsey Cattle Club is formed at the home of Augustus Ward in Farmington, Connecticut; considering that Guernseys only came to this country forty-five years ago, they're not doing badly at all. Guernsey cows note that American Jerseys have had their own club for nine years; it was started by forty-three dairy-men in Newport, Rhode Island.

An association is founded in the state of Ohio with the expressed purpose of preventing cruelty to livestock while they are being transported. Cows are touched by this unexpected concern. They also hope that someone will broach the subject of railway stockcar design as it applies to cattle; cows would like to see much wider spaces between the sideboards on cattle cars. Right now, the cars are very crowded and stuffy, and it's difficult to get a proper view of the countryside.

A small number of west African N'Dama cattle are present at the Paris World Exhibition. A news-paper account describes the N'Dama as entirely humpless, long and slender horned, small in stat-ure, sand colored, and unquestionably of pure *primigenius* type. There is also considerable discussion of the origins of the breed. Some people suggest that they may be a dwarf version of the Egyptian longhorn. N'Dama cows enjoy the attention but are not accustomed to having people look into their backgrounds.

Joseph Farwell Glidden of De Kalb, Illinois, invents barbed wire and begins manufacturing it almost immediately. Cows are not the least bit impressed with this invention. They think Joe could have put his talents to much better use.

The first American cheese factory is built in Oneida County, New York. Within twenty years two thirds of all the cheese produced in America is factory made. Cows in England soon hear farmers grumbling about foreign competition. The next thing you know a British cheese factory is opened with great fanfare in the county of Derbyshire.

Pierre Auguste Renoir, a well-known painter of sensuous nudes, completes a canvas entitled *The Return from the Fields;* Renoir's painting depicts a young dairymaid with a beautiful French cow. The following year Alfred Philippe Roll paints *Farmer's Wife,* a large painting with girl, milk pail, and cow. French cows believe both these artists are on the right track.

The real cultural event of the century takes place when a Scottish writer publishes a poem entitled "The Cow." Cows first hear Robert Louis Stevenson's poem recited by a young schoolgirl; nothing has ever spread throughout their world more quickly.

> The friendly cow, all red and white,
> I love with all my heart:
> She gives me cream with all her might,
> To eat with apple-tart.
>
> She wanders lowing here and there,
> And yet she cannot stray,
> All in the pleasant open air,
> The pleasant light of day;
>
> And blown by all the winds that pass,
> And wet with all the showers,
> She walks among the meadow grass
> And eats the meadow flowers.

There is great excitement among cows in Belgium over a new musical instrument called a saxophone. Cows hear it said that young Adolphe Sax is extremely clever; it is also suggested that he should spend more time at home and less in France.

Cows on the island of Guernsey are pleased when French poet Victor Hugo decides to spend his time in exile on their island. They are particularly impressed when they hear that he has written a book which draws attention to the hardships of nineteenth-century children. Cows think it's disgraceful that children are expected to keep the wheels of industry turning.

Cows in continental Europe are starting to notice improvements in transportation. Aside from railway networks, travel on inland waterways becomes important; extensive canal systems are built in France and Germany. Most significant of all, manufactured goods are now moving about in an atmosphere of free trade. Cows believe that as long as people are trading there's less chance of them fighting.

The name Sigmund Freud is now heard with some regularity. The Austrian psychoanalyst apparently believes that mental disorders are often the result of suppressed sexual desire. Whether this is true or not, Dr. Freud is arousing great controversy. Cows also notice a direct relationship between the doctor's fame and the demand for good-quality cowhide—it's being used to cover special couches used in Dr. Freud's type of work.

A huge fire in Chicago claims an estimated two hundred and fifty lives and destroys millions of dollars' worth of property. Newspaper reports openly suggest that the city's worst fire was started when a cow belonging to Mrs. Patrick O'Leary kicked over a lantern while being milked. Cows are not convinced that this is what really happened, but if it did it was certainly an accident.

Nicklaus Gerber in Switzerland and Stephen Babcock in the United States develop the first practical methods for determining the amount of butterfat in milk. Cows understand for the first time the true value of butterfat—more butterfat means that farmers can charge more for their milk. It also means that farmers are now prepared to feed their cows more carefully.

The first dairy school of collegiate rank is opened in the state of Wisconsin. Cows are initially disappointed when only two students attend the first year's classes but are reassured when seventy students enrol in the second year. These young people are smart.

Its hard to believe that two poets would be inspired to celebrate the cow in one century, but this is exactly what happens. American humorist Gelett Burgess publishes a short nonsensical poem entitled "The Purple Cow":

> I never saw a Purple Cow,
> I never hope to see one;
> But I can tell you, anyhow,
> I'd rather see than be one!

Some years later, Mr. Burgess, tired of being identified as the author of this poem, writes:

> Ah, yes, I wrote the "Purple Cow"—
> I'm sorry, now, I wrote it!
> But I can tell you, anyhow,
> I'll kill you if you quote it.

As the nineteenth century comes to a close, there is a strong feeling of optimism in the cow world, a sense of what people can accomplish when they put their minds to it. Surely with all the advances that have been made in transportation and science and medicine, it's not unreasonable to assume that the next century will bring even greater achievement. Maybe things will just keep getting better.

Twentieth Century

Cows are beginning to notice a lot more travel among heads of state. Prince Ito of Japan travels to Russia to negotiate with Emperor Nicholas; British King Edward VII no sooner returns from Paris when French President Emile Loubet visits him in London; President Theodore Roosevelt visits the Panama Canal, becoming the first American President to travel outside the country on official business.

There's a noticeable increase in the number of assassinations. Hardly a year goes by without a king or a premier or a minister of government getting gunned down. In the first quarter of this century cows hear of at least twenty-four assassinations and one very close call.

Countless new states are forming throughout the world. Australia and New Zealand become independent dominions within the British Empire; Cuba becomes a United States protectorate; Norway separates from Sweden; the Union of South Africa is formed; republics are proclaimed in Poland, Austria, Turkey, Albania, and Lebanon. Everywhere cows turn, people seem to be doing things with nationalistic fervor.

Two United States aviators, Orville and Wilbur Wright, successfully fly a powered airplane near Kitty Hawk, North Carolina. Cows are amazed when this spectacular achievement fails to excite the American public. Perhaps the Wright brothers should have timed their event more carefully; with Christmas only eight days away, people are busy with last-minute shopping.

The Herringbone milking parlor comes into vogue in Australia. This means that cows can no longer expect to be milked where they happen to be standing; instead they'll be required to file into a barn especially designed for milking efficiency. Australian cows are confused by the term herringbone—none of them has the slightest notion what a herring bone looks like.

After forty-four years of continuous peace among European states, cows are beginning to entertain the notion that wars, at least major ones, are a thing of the past—but sadly this proves not to be the case. The Great War of 1914–1918 involves a total of sixteen nations from every continent of the civilized world. It is a war of unprecedented scale, where huge industrialized societies and vast resources are mobilized against each other—the greatest example of misdirected energy cows have ever witnessed. By war's end, ten million people have died in battle, and a further twenty million are missing or wounded.

British-Italian engineer Guglielmo Marconi discovers a way to send messages without the use of wires. Shortly after the first radio sets appear, the first transcontinental telephone call is made in the United States, and a wireless service is established between America and Japan. Cows wonder if all these new ways of communicating are going to have much of an effect on the way people relate to one another.

A new type of music is developing in the southern United States. Cows hear jazz for the first time when people from New Orleans begin playing it on country outings. It's also heard at funerals and weddings. While many people seem to consider the rhythms of jazz a bit too boisterous, cows take to it almost immediately.

With well over four million radio sets now in use, music is being heard just about everywhere. Not all of this is good listening, of course, but cows appreciate the variety. "Barney Google" becomes a hit song in America. Another favorite is Irving King's "Show Me the Way to Go Home"; cows notice that Mr. King's song is usually sung late at night and quite often by men who appear to be celebrating.

The twentieth century brings a noticeable change to the art of dancing; people in industrialized societies are starting to loosen up. In America the foxtrot and Charleston are followed by the elegant tango. If cows were able to dance, the tango would be their first choice.

Some of the most promising developments in the twentieth century are directly related to women. In Austria, Britain, the United States, and Canada women win the right to vote. The Americans set aside a special day to honor mothers. In Britain, the Girl Guides are established, and Lady Astor is elected Member of Parliament. In Germany, women are admitted to universities for the first time. Cows are convinced that the world would be much better off if some of these things had come about sooner.

Revolution breaks out in the Russian city of Petrograd. Within three years a new political order has been established. In the process, thirteen million people have perished in war and famine. Cows initially find it difficult to sort out what's going on but soon learn that the new Union of Soviet Socialist Republics plans to eliminate the capitalist system entirely; its economy will be based on the communal ownership of all property. Cows can appreciate the thinking behind this approach but have serious doubts about a Marxist society ever working.

In Italy Benito Mussolini comes to power as a fascist dictator. As cows understand it, fascism rejects the idea of individual freedom and emphasizes racial superiority—hardly the kind of thing this world needs.

Heinz and Lutz Heck, two brothers working in Germany, are determined to reconstitute the wild aurochs. Heinz, director of the Munich Zoo, crossbreeds Hungarian and Podolian steppe cattle, Scottish Highland and Alpine breeds, Frisians and Corsicans. Lutz, working at the Zoological Gardens in Berlin, starts with Spanish fighting cattle, the Camargue breed from the lower Rhône, Corsican, and English Park cattle. After some years both zoologists succeed in producing animals remarkably similar to their ancestoral type. Cows in Germany are not entirely sure that these experiments represent progress; as well as being physically similar to their ancient predecessors many of these reconstituted animals are extremely fierce and incredibly cranky.

Women on both sides of the Atlantic continue to make great strides. An American swimmer, Gertrude Ederle, becomes the first woman to swim the English Channel. In Britain, female suffrage is reduced from age thirty to twenty-one. In the United States, Amelia Earhart flies solo across the Atlantic; a few years later, Mrs. Hattie Caraway becomes the first woman elected to the U.S. Senate.

British women are the focus of two controversial books: *The Intelligent Woman's Guide to Socialism and Capitalism,* by George Bernard Shaw, and *Lady Chatterley's Lover,* by D. H. Lawrence. Although Mr. Lawrence's book is not readily available, it manages to cause a public outcry. Cows in Nottinghamshire are not surprised. David Herbert Lawrence created a great scandal some years back when he ran away with a local professor's wife; people in this part of the world talk about it as if it happened yesterday.

There's considerable fuss in the cow world when it's learned that bulls will no longer be required to provide their services on a personal basis. The new approach in cattle breeding is known as artificial insemination; by using frozen semen, as many as one hundred thousand cows can now be impregnated without so much as an introduction. Cows think this is a classical example of the twentieth century going too far.

The American composer Irving Berlin writes

a piece of music which cows immediately adopt as their own personal theme song. "Blue Skies" is but two years old when Bud De Sylva and Lew Brown write "You're the Cream in My Coffee." Next, cows in America hear "I'm an Old Cowhand from the Rio Grande," followed by "It's a Big, Wide, Wonderful World." While all of these songs are considered special in the cow world, none of them express bovine concerns more successfully than "Don't Fence Me In." This is a song cows could easily have written themselves.

Another American, Ogden Nash, publishes a small book of poems entitled *Verses from 1929 On.* Cows discover that Mr. Nash has been kind enough to include a short poem about them:

> The cow is of the bovine ilk;
> One end is moo, the other, milk.

Cows like this poem a great deal but find it a bit short.

The first cow ever flown in an airplane goes aloft with a group of highly spirited reporters on February 18, 1930. Elm Farm Ollie, a well-behaved Guernsey, is not only milked during the flight, but her milk is sealed in paper containers and parachuted over the city of St. Louis, Missouri. Cows consider this one of the most spectacular events in their entire history; in truth, they've wanted to fly from the moment they first watched the Chinese flying kites in 1000 B.C.

The first African cattle ever to set hoof on American soil disembark at New York City following a month of cramped oceanic travel. Sixteen Africander bulls and thirteen cows are then held in Clifton, New Jersey, for a total of eighty-eight days. This is called living in quarantine. Finally all twenty-nine animals arrive at their new home in Kingsville, Texas. News of this event creates a great stir in the American cow world.

The first drive-in movie theater opens in a ten-acre plot near Camden, New Jersey, and the first movie neighboring cows get to see is *She Done Him Wrong,* starring Mae West. It is duly noted that not everyone at the drive-in is interested in what's on the screen.

Milk wars are being waged between farmers and large dairy processors in the eastern United States; to help alleviate some of the bad feelings, Elsie the Cow is created by the Borden Milk Company. At first, Elsie appears in print media only, but response from the American public is so enthusiastic, the decision is made to select a real live cow to represent her. A seven-year-old blue-blooded Jersey from Brookfield, Massachusetts, whose registered name is You'll Do Lobelia, is the lucky bovine. In the next few decades Elsie and her family become the toasts of America. Cows almost burst with pride.

Joseph Stalin becomes supreme dictator of the largest country in the world. He immediately institutes a series of five-year plans designed to bring about the restructuring of Soviet industry; this involves the creation of collective farms. Few people are impressed with this new approach, least of all those who are forced to do the work, but then Stalin is not concerned about popularity. With unspeakable brutality he eliminates anyone who dares to oppose him; during a single decade of terror, twenty million people are ruthlessly slain.

With assistance from Adolf Hitler and Benito Mussolini, a fascist dictatorship is firmly established in Spain under Francisco Franco. Within three years, the civil war which brings Franco to power has taken the lives of seven hundred and fifty thousand people.

World War II begins in Europe. Six years later, fifteen million soldiers and thirty-five million unarmed civilians are dead; twenty-seven countries have taken part in the biggest war ever waged.

In Germany and Poland, cattle cars are used to transport Jews, gypsies, and other minority groups to camps where they are held without trial. At Auschwitz, Treblinka, Belsen, Buchenwald, and Dachau millions of people, including children, are gassed or die of disease or starvation. Two thirds of European Jewry are exterminated in a Nazi attempt to rid the world of all non-Aryan peoples. At times like this, cows would prefer not to be a part of the human story at all.

In the Pacific, U.S. President Truman orders the first use of an atomic bomb, dropped on the city of Hiroshima; three days later a second bomb is dropped on Nagasaki. In total, over two hundred thousand people are killed or injured. The plane carrying the Hiroshima bomb was piloted by an Iowa farm boy named Paul Tibbets. Cows learn, much to their dismay, that Mr. Tibbet's Boeing B-29 is named the *Enola Gay* after his mother's cow.

English writer George Orwell creates a furor on both sides of the Atlantic with a new book called *Animal Farm*. Judging from what cows hear, this is not the kind of farm they would enjoy— the pigs are said to be very unfriendly and extremely manipulative.

Cows in America are beginning to notice that fewer children are playing out of doors in early evening; at the same time a bluish light is seen flickering in the windows of many homes. Televisions have suddenly become the rage; in 1950 alone, fourteen million television sets are sold.

Cows have nothing against sports in general but it does occur to them that golf is getting out of hand; there are now five thousand golf courses taking up a million and a half acres of prime land. This wouldn't be so bad except good pastureland is getting scarcer every year.

Cows in India continue to be revered; all but two states have enacted laws strictly forbidding their mistreatment or slaughter. Special nursing homes called Gowshallas have been established to take care of animals that have become barren or too feeble to work. Twice yearly, cow festivals are held throughout the country, one in late summer after the rains and a second following the harvest. Needless to say, Indian cows enjoy these occasions—aside from being decorated with garlands, they're fed on succulent mustard greens and sprinkled with holy water.

Indian farmers are sometimes accused of making too much fuss over their cows, but people who hold this view have obviously not taken the time to consider all the facts. To begin with, the country's two hundred million cattle do two thirds of the work on Indian farms. They also produce eight hundred million tons of manure annually, much of which gets plowed back into the soil with the balance being used as cooking fuel. People should take these factors into account before making their sweeping comments about cow worship.

Cows in Japan are amazed to learn that in addition to being fed on beer, beef cattle are now being massaged on a regular basis. All of this special attention is the result of a nationwide demand for Kobe beef. Japanese cows would like someone to start the rumor that milk is more nutritious when cows are given the Kobe treatment.

There are now one billion cattle in the world producing an estimated eighty-eight billion pounds of beef annually. Perhaps even more impressive are the world figures for milk production. It is estimated that cows produce seven hundred billion pounds per year, enough milk to provide each man, woman, and child in the world with one half pint of milk daily. Unfortunately, milk production is anything but uniform; eighty-five percent of the total supply is produced in an area

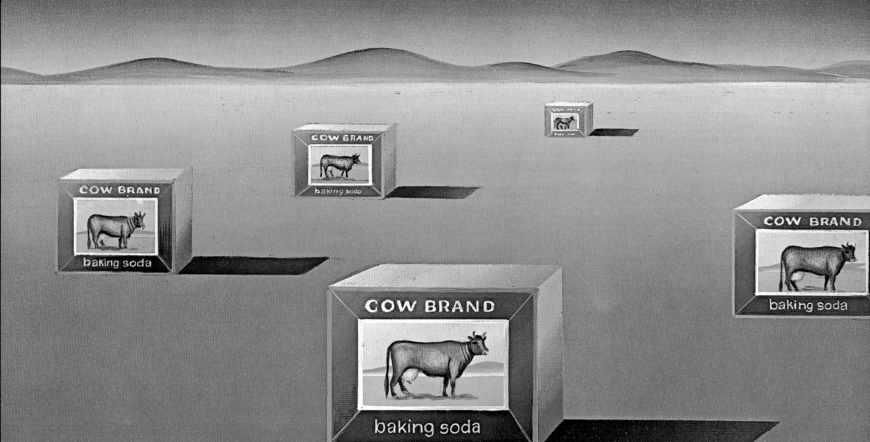

alaingauthier

supporting only a third of the world's population.

There's a considerable flurry among cows in Ohio when an eight-year-old Holstein establishes a new world record for milk production. Breezewood Patsy Bar Pontiac of Vienna, Ohio, produced forty-five thousand two hundred and eighty pounds of milk in just three hundred and sixty-five days. Cows are suitably impressed; they're also concerned that farmers will now expect this kind of production from everyone.

The dairy world has barely adjusted to Breezewood's remarkable achievement when a cow belonging to Harold Beecher of Rochester, Indiana, outdoes the Ohio cow by ten thousand three hundred and eighty pounds. Holstein cow Beecher Arlinda Ellen creates a stir as far east as New York City—her story appears in a Sunday edition of *The New York Times*. As far as cows know, this is a first.

Michigan cows are stricken with grief when a chemical poison known as PBB is accidentally fed to farm animals throughout the state. To prevent a public health hazard, twenty-seven thousand five hundred and eighty cattle are destroyed by state officials. Eighty percent of this number are cows; in addition, four thousand six hundred swine, one thousand five hundred sheep, one and a half million poultry, thirty-two rabbits, two goats, and two horses are also put down. With vast amounts of herbicides and pesticides now in use, plus millions of tons of industrial waste spilling into the air and water, it's becoming increasingly difficult for farm animals to know what to eat. Cows notice that while grasslands may appear lush and green, the grass doesn't always taste so good.

Cows once again are making a direct contribution to medical science. In a new approach to vascular surgery developed in the United States, doctors are using arteries from the necks of cows to replace diseased blood vessels in humans. Cows are pleased to discover that arterial grafts are particularly useful in the treatment of arteriosclerosis, a disease occurring among the elderly.

Embryo transplants are now common among highly prized cows. After receiving fertility hormones to make them superovulate, champion cows are artificially inseminated with sperm from a prize bull. Five days later, an operation is performed to remove the embryos so they can be transplanted into a low-grade surrogate mother. In this way, twentieth-century cows not only get impregnated by a bull they never meet—they produce calves they will never carry.

Word reaches the cow world that combined military spending is now in excess of four hundred billion dollars a year; that more than a billion dollars is spent every day to insure that wars will have the benefit of man's latest technologies. Some developing countries spend three times as much on arms as they do on health care. It is also noted that nuclear missiles are often buried in the gentle earth away from public view. Would-be pastureland is not always what people think it is. Cows wish that people would finally put an end to all this insanity.

It is now clear that *Bos primigenius* cannot continue to follow world events as they have done in the past. After lengthy and difficult discussions a unanimous decision is reached. Henceforth all cows will go on about their business without feeling obliged to keep track of everything that's going on around them. Before making this move however, every effort will be made to collect their thoughts and pass them on in written form. The method by which cow notes are actually written down was not arrived at easily and while cows suspect that people will be curious about how this is done, it will have to remain a closely guarded secret.

We address this letter to all of you and hope that whoever finds this milk bottle will be kind enough to pass our message on. We tried to keep our observations brief. This of course, has meant passing over many points of interest. We would have enjoyed, for example, paying much more attention to architecture and music and poetry. We would have liked to have had more time to describe exotic cattle breeds and the customs of those who care for them. For brevity's sake we have had to eliminate a great deal—we are sure you will understand.

Dear People:

We are leaving you. All female members of genus *Bos,* family Bovidae are retreating to the background, to a quieter, less stressful life. We will no longer act as observers of the human scene—from now on we will live in our pastures enjoying whatever music flows our way.

It's not easy for us to make this announcement, but with life as hectic as it is, with cows having less and less access to what's going on, we have no other choice. We must say, however, that *Bos primigenius* have no intention of deserting you entirely; we will do our utmost to continue providing you with all the products and services that are now a part of bovine legacy—as always we will do this without complaint.

As faithful servants we have nourished you for thousands of years—when you first lived in caves, when you moved from caves to tents made from our skins and bones. We watched your tents form clusters, until clusters grew into villages and villages became towns. We were excited about your first cities; we watched them grow into city-states. We witnessed one empire after another, until finally vast civilizations were formed. We fed you continuously and never complained; we pulled your clumsy plows and put up with all kinds of mistreatment.

We watched you fight and tear each other apart, but we also watched you dance. At times, we watched you achieve incredible things.

As simple cows, we have given you everything we have. With hides off our backs you have made shoes and belts and handbags; you sit on leather sofas, carry leather luggage, and play baseball with our leather mitts. At one time our hooves gave you buttons; our horns were used to make music and to store your gunpowder and ink. With casein from our milk you made paint. We have given you so much milk, cheese, butter, and ice cream it would be impossible for you to calculate the amounts. With our help you discovered vaccines to prevent sickness and produced insulin for diabetics. We provide much of the protein on this planet. Because of us, millions upon millions of human beings are not only fed but have jobs in all of the industries which would not exist without us. There is little left for us to give.

Before we go, we want to wish you all the very best and we want you to know that we believe in you. We believe that you can make the right decisions, but you must want to make them. You can continue to build sinister weapons, or you can take all that amazing energy and redirect it; you can create a decent world free from war or the threat of war. You can allow the human spirit to soar where it has never been.

Just one more thing. When you see us in our pastures staring blankly into space or appearing withdrawn, you will probably think of us as stupid looking, if you think of us at all. Take a closer look—you will see bovine serenity; look closer still and you might notice that some of us are soaring. Again our best.   Cows

ACKNOWLEDGMENTS

Cows would like to express their gratitude to the many people who assisted the author in the preparation of this book. We know how busy everyone is nowadays and we realize that it's not always easy to give freely of one's time. Writing our story required a great deal of travel on the author's part and a lot of support was needed along the way. Cows are especially grateful to the following people.

In Belgium: Mr. and Mrs. Jean Dierickx.

In Canada: David Baxendale, David Bergmark, Dr. Edward Burnside, Michael Dalton, Dianne Davy, Bill Hanrahan, Jim Hornby, Christopher Hyde, Rodney Jones, Harry and Lori Kennedy, G. S. MacKay, Sister Vera Mac-Lellan, Andrew and Marion Mikita, Libby Oughton, Jean Perry, Leonard and Audrey Poetschke, Bella Pomer, Carl Sentner, Doug Tower, Harold Verge, Christopher Wells, Kennedy Wells, Robin Esmond-White.

In Czechoslovakia: Josef Paleček, Petr Sis.

In Egypt: Helmi Mourad.

In Finland: Jukka Veistola.

In France: Daniele Bour, André Dahan, David Douglas Duncan, Ezdin Hosni, Dimitri Savitizy, Gerard and Susan Serviant.

In Great Britain: Cows will always remember the generosity of Nicholas Dawe and his associates at Folio. Cows are also grateful to Dr. Juliet Clutton-Brock, John Friend, Dr. Alan Gentry, Robin Hill, Keith Lilley, Bruce and Dorothy Mickel, Deborah Pate, Tom and Jane Thomson, Laurel Wade, Ray Winder.
  Guernsey: Rhona Cole, Nigel Jee.
  Jersey: Derrick Frigot, Brian Skelly, Joan Stevens, Dorothy Wallbridge.

In Greece: Costas Lapidis (Crete).

In Germany: H. Sigfrid Kúbe, Dejan Marinkovic, Rudi Seitz.

In India: G. L. Joshi, Vishnu and Rita Mathur, Jaya Rastogi, Ramesh Sanzgiri.

In Italy: Gino Brocato, Stewart and Judy Campbell, Lisa Clark, Bob and Marie Sharp.

In the Netherlands: H. de Boer (Friesland), Heinz Edelmann, Ruth Gembala, Harrie Leyten, Hans Mohs, Gerald and Lidia Postma, Maaike Sigar.

In Poland: Olga Siemaszko, Waldemar Świerzy, Marcin Wyszomirski.

In Switzerland: Anna Marie Germann, Rolf Inhauser, Trudi Marty, Hans Sauerländer, Denis Robert-Tissot.

In Tanzania: Hashim Sineni.

In the United States: Cows are particularly indebted to Robert Gottlieb, Editor-in-Chief of Alfred A. Knopf, for recognizing that our time has finally come and to Martha Kaplan for her patience and good humor while sorting out our notes. Special thanks are also due Carol Brown Janeway and Marion Bundy in Editorial and Dennis Dwyer, Joe Marc Freedman, and John Woodside in Production and Design. In addition we are grateful to Karen Robinson, Judy Loeser, and Bob Scudellari at Random House and to the following people—Michael Botwinick, George Castellani, Seymour Chwast, Erika Faisst, Milton Glaser, Paul Gottlieb, Sean Kelly, Bea Losito, Patricia Luca, Mary Maguire, Christopher Manning, Robert Priest, Frederic Roy, Alan Sparger, Jr., Mack and Barbara Stanley, Lita Telerico, Paula Vaught, Jim Yanizyn, Barry Zaid, Philip Zinke.

In Yugoslavia: Ivan Generalić, Ksenija Gregur, Zoran Krzisnuk, Seka Prebeg, M. Sôla.

A NOTE ON THE TYPE

The text of this book was set via computer-driven cathode-ray tube in a modernized version of Cheltenham Old Style, originally designed by the architect Bertram Grosvenor Goodhue in collaboration with Ingalls Kimball of The Cheltenham Press of New York. Cheltenham was introduced in the early twentieth century, a period of remarkable achievement in type design. The idea of creating a "family" of types by making variations on the basic type design was originated by Goodhue and Kimball in the design of the Cheltenham series.

The display typeface, Gallant Bovina, was created especially for this book by the author. Based on calligraphic models it combines soft italic contours with the fluidity of cows rushing home just before a rainstorm.

Composed by Centennial Graphics, Inc., Ephrata, Pennsylvania
Printed and bound by Officine Grafiche di Verona, Arnoldo Mondadori Editore, Verona, Italy
Art direction and design by Marc Gallant

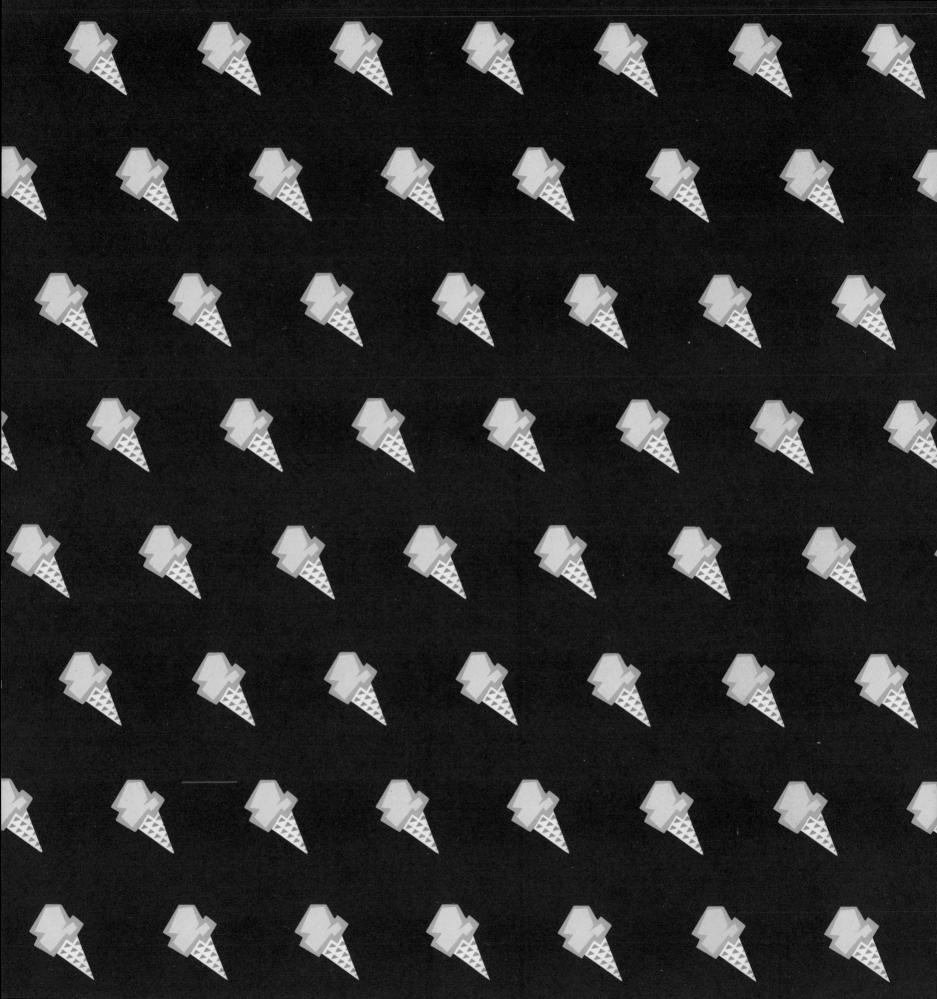